I0475804

EPA's Information Systems and Data Are at Risk Due to Insufficient Training of Personnel With Significant Information Security Responsibilities

Report No. 14-P-0142   March 21, 2014

Abbreviations

CIO	Chief Information Officer
EPA	U.S. Environmental Protection Agency
FISMA	Federal Information Security Management Act
GAO	Government Accountability Office
IMO	Information Management Officer
IPA	Independent Public Accounting
ISO	Information Security Officer
ISSO	Information System Security Officer
IT	Information Technology
NIST	National Institute of Standards and Technology
OEI	Office of Environmental Information
OIG	Office of Inspector General
OPM	Office of Personnel Management
SAISO	Senior Agency Information Security Officer
SIO	Senior Information Officer
SP	Special Publication

At a Glance

Why We Did This Review

The U.S. Environmental Protection Agency's (EPA's) Office of Inspector General (OIG) contracted with KPMG LLP, an independent public accounting (IPA) firm, to conduct an audit of the qualifications and current skills of EPA personnel with significant information security responsibilities, to determine their training needs and evaluate consistency with the E-Government Act of 2002.

The E-Government Act requires federal agency information technology (IT) security personnel to maintain sufficient training and knowledge to conduct their duties.

This report addresses the following EPA theme:

- *Strengthening EPA's workforce and capabilities.*

For further information, contact our public affairs office at (202) 566-2391.

The full report is at:
www.epa.gov/oig/reports/2014/20140321-14-P-0142.pdf

EPA's Information Systems and Data Are at Risk Due to Insufficient Training of Personnel With Significant Information Security Responsibilities

What the IPA Auditor Found

The EPA lacks an information security role-based training program that defines specific training requirements for personnel with significant information security responsibilities. Implementation of the EPA's information security training program is hindered by inconsistent assignment of information security roles across the various EPA offices. The current training program does not consider specific needs of technical and managerial personnel responsibilities for implementing information security as required by the federal guidance. Management has not completed efforts to tailor the existing training programs to align it with the EPA's organizational structure. The EPA's decentralized structure creates differing levels of information security implementation and oversight of training requirements. As a result, training may be insufficient to assure management that personnel with significant information security duties have the skills and understanding necessary to identify, prevent or mitigate vulnerabilities affecting the EPA's information systems and infrastructure.

The EPA places its information systems and data at risk due to an organizational structure that has not specified required duties and responsibilities to ensure personnel are trained on key information security roles.

The IPA is responsible for the content of the audit report. The OIG performed the procedures necessary to obtain reasonable assurance about the IPA's independence, qualifications, technical approach and audit results. Having done so, the OIG accepts the IPA's conclusions and recommendations.

Recommendations and Agency Corrective Actions

The IPA's report recommends that the Assistant Administrator for Environmental Information: (1) define key information security aspects and duties for each security role; (2) provide additional training options specific to the federal information security environment and EPA information security roles; (3) standardize the terminology and definition of responsibilities for key IT security roles; and (4) provide clearer delineation of which EPA organizations should be responsible for delivering specific elements of information security role-based training. EPA agreed with the recommendations and is taking corrective action.

Noteworthy Achievements

The EPA conducts an annual Security and Operations conference. The EPA also implemented an annual specialized training requirement for employees with significant information security responsibilities.

UNITED STATES ENVIRONMENTAL PROTECTION AGENCY
WASHINGTON, D.C. 20460

March 21, 2014

MEMORANDUM

SUBJECT: EPA's Information Systems and Data Are at Risk Due to Insufficient Training of
Personnel With Significant Information Security Responsibilities
Report No. 14-P-0142

FROM: Arthur A. Elkins Jr

TO: Renee P. Wynn, Acting Assistant Administrator and Chief Information Officer
Office of Environmental Information

The independent public accounting (IPA) firm KPMG LLP conducted this audit on behalf of the
U.S. Environmental Protection Agency's (EPA's) Office of Inspector General (OIG). This is the IPA's
report on the subject audit conducted on behalf of the OIG. This report contains findings that describe
the problems the IPA identified and corrective actions the IPA recommends. The Senior Agency
Information Security Officer is the primary official responsible for the agency program that KPMG
reviewed during this audit. This report represents the opinion of the IPA and does not necessarily
represent the final EPA position. The agency concurred with all the report's recommendations and
provided high-level planned corrective actions with milestone dates, which KPMG found acceptable.

Action Required

Based upon your response to the draft report, we will close this report in our audit tracking system upon
issuance. We believe the proposed actions, when implemented, will adequately address the report's
findings and recommendations. Please provide updated information in the EPA's Management Audit
Tracking System as you complete each planned corrective action or revise any corrective actions and/or
milestone dates. If you are unable to meet your planned milestones, or believe other corrective actions
are warranted, please send us a memorandum stating why you are revising the milestones or why you
are proposing alternative corrective actions, as required by EPA Manual 2750.

If you or your staff have any questions regarding this report, please contact Richard Eyermann,
acting Assistant Inspector General for Audit, at (202) 566-0565 or eyermann.richard@epa.gov;
or Rudolph M. Brevard, Director, Information Resources Management Audits, at (202) 566-0893 or
brevard.rudy@epa.gov.

March 4, 2014

Re: Assessment of EPA Personnel with Significant Information Security Responsibilities

Thru: Arthur A. Elkins, Jr.
        Inspector General

To: Renee P. Wynn, Acting Assistant Administrator for
        Environmental Information and Chief Information Officer

Thank you for providing KPMG LLP (KPMG) with the opportunity to assist the Environmental Protection Agency Office of Inspector General in performing an assessment of EPA personnel with significant information security responsibilities.

In summary, we found opportunities for improvement in the development and implementation of EPA's role-based information security awareness program across the regional and program offices supporting the EPA's mission. Although Federal personnel generally demonstrated a high level of awareness of responsibilities associated with their assigned information security roles, EPA could take additional steps to formalize the management of roles and responsibilities across its workforce and align training requirements to those roles.

Please provide your written comments to the EPA OIG points of contact.

Sincerely,

Tony Hubbard

Tony Hubbard, Principal

EPA's Information Systems and Data Are at Risk
Due to Insufficient Training of Personnel With
Significant Information Security Responsibilities

14-P-0142

Table of Contents

Chapters

1 Introduction 1

    Purpose 1
    Background 1
    Responsible Office 1
    Noteworthy Achievements 2
    Scope and Methodology 2

2 Information Security Role-Based Training Efforts Can Be Better Defined 3

    Definition of Roles and Responsibilities Is Incomplete or Inconsistent 3
    Role-Based Training Is Not Specific to Assigned Information Security Roles ... 4
    Available Skillport Training Does Not Align With EPA Professionals' Needs 5
    Recommendations 7
    Agency Comments and KPMG Evaluation 8

3 Information Security Governance Supporting Training Efforts
Can Be Improved 9

    Assignment of Information Security Roles Is Inconsistent 9
    Organization Structure for Information Security Roles Is Inconsistent 9
    Recommendations 12
    Agency Comments and KPMG Evaluation 12

Status of Recommendations and Potential Monetary Benefits 13

Appendices

A   Survey Questions 14

B   Agency Response to Draft Report 17

C   Revised Agency Corrective Action Plan to Report Recommendations 20

D   Distribution 22

Chapter 1
Introduction

Purpose

The objective of this review was to evaluate the qualifications and current skills of U.S. Environmental Protection Agency (EPA) personnel with significant information security responsibilities, determine their training needs, and determine whether the EPA's information security workforce possesses the knowledge, competencies and skills necessary to meet agency goals as mandated by the E-Government Act of 2002.

Background

On December 17, 2002, the President signed into law the E-Government Act of 2002, providing a comprehensive framework for information security standards and programs. Title III of the E-Government Act is the Federal Information Security Management Act (FISMA), which requires federal Chief Information Officers (CIOs) to assess and report on the status of their agency's information security program. FISMA focuses on information security program management, implementation and evaluation aspects of the security of federal information systems. FISMA also codifies existing guidance from the Office of Management and Budget and National Institute of Standards and Technology (NIST), as well as regulations from the Clinger-Cohen Act of 1996. FISMA requires that information security personnel possess professional qualifications, including training and experience, required to administer the functions described in the act. In addition, FISMA requires federal agencies to adequately train all personnel with significant information security responsibilities. According to FISMA, an agency must ensure that it has trained personnel sufficient to assist the agency in complying with the requirements of the act.

Responsible Office

EPA's Chief Information Officer (CIO), within the Office of Environmental Information, is responsible for developing and maintaining an information security program as required by the E-Government Act of 2002, Title III Information Security, also known as the Federal Information Security Management Act. Within the CIO office, EPA's Senior Agency Information Security Officer (SAISO) is responsible for developing and maintaining role based training, education and credentialing requirements to ensure personnel with significant information security responsibilities receive adequate training with respect to personnel's responsibilities.

Noteworthy Achievements

We generally noted that the EPA information security personnel we interacted with through the survey and interviews did appear to possess the qualifications, skills and experience necessary to execute their assigned security-related responsibilities. We also noted that the EPA has some processes in place to promote information security training and knowledge, such as offering an annual Information Security Officer (ISO) conference through classroom and webcast-based form, conducting monthly ISO coordination meetings, and implementing an annual specialized training requirement for employees with significant information security responsibilities under the direction of the SAISO.

Scope and Methodology

We initially conducted a Web-based survey to gather anonymous input on improving the EPA's information security program (356 EPA federal employees completed the survey). We applied the survey results to gather data related to the responders' office/region, extent of security training, years of relevant experience, relevant professional certifications, operational responsibilities, and familiarity with the EPA's information security policies and practices. We have provided the survey questions as Appendix A to this report.

We also interviewed 87 EPA employees with significant information security responsibilities located at the following offices:

- Headquarters program offices.
- Research Triangle Park (North Carolina).
- Region 5 (Chicago, Illinois).
- Region 8 (Denver, Colorado).

The results of our survey and interviews are not statistically valid, and therefore we cannot project the results to the EPA organization as a whole.

We also reviewed EPA policies and procedures relevant to this review.

We initiated the review in March 2012, conducted the survey from October to December 2012, and completed the review procedures in August 2013. Evaluation fieldwork was conducted in accordance with generally accepted government auditing standards.

Chapter 2
Information Security Role-Based Training Efforts Can Be Better Defined

The EPA's information security training and awareness programs are based upon inconsistent assignment and definition of related roles and responsibilities across the various regional and program offices. Existing training materials do not adequately consider specific needs of technical and managerial personnel with assigned and collateral responsibilities for information security as required by the EPA, NIST, and Office of Personnel Management (OPM) requirements. Management has not completed efforts to tailor existing training programs to align with the EPA's organizational structure. As a result, training may be insufficient to assure management that information security personnel have the skills and understanding necessary to identify and prevent or mitigate the threat of vulnerabilities affecting the EPA's information systems and infrastructure and underlying financial or mission-critical business data.

Definition of Roles and Responsibilities Is Incomplete or Inconsistent

There are no specific information security training requirements or curriculum defined for personnel with significant information security roles, such as an ISO. In the memorandum, "Training for EPA Employees with Significant Information Security Responsibilities," addressed to ISOs (initially issued on June 23, 2003, and reissued each subsequent fiscal year), the SAISO listed 17 information security roles as a basis for identifying individuals in the ISOs' organizations subject to role-based training requirements. The 17 defined roles are generally comparable to roles defined in NIST guidance, but the responsibilities of the roles can vary greatly. For instance, many EPA information security professionals are called ISOs, but the individuals assigned the role of ISO can have widely varying duties and levels of responsibilities. They may be a Primary ISO for an office or region, a local ISO at a field location, an ISO for a single system, or a security professional at a data center who supports network or systems infrastructure.

Further, there are inconsistencies in the naming and definition of information security roles among various EPA policies and organizations. For example, EPA CIO Policy 2150.3, *EPA Information Security Policy*, defines the following information security roles: CIO, SAISO, Senior Information Officer (SIO), Authorizing Official's Designated Representative, System Owner, Information System Security Officer (ISSO), and Common Controls Provider. However, CIO Procedure 2150.3-P-02.1, *Information Security – Interim Awareness and Training Procedures v.3.1*, lists roles and responsibilities for the CIO, System Owners, Information Owners, Information Technology (IT) Security Program Managers, managers and supervisors, EPA Administrator, and general-end users. Although there are references to other roles, the procedure does not define roles and responsibilities with respect to the EPA's information security training and awareness program for other defined positions within the agency, including the SAISO, SIOs, Primary ISOs, and ISSOs.

Role-Based Training Is Not Specific to Assigned Information Security Roles

Of the 356 survey respondents, 51 (14 percent) indicated they did not receive specialized role-based information security training in the prior year. Specific training is needed for many information security roles, such as: 1) NIST defined information security roles, such as those supporting system certification and accreditation and continuous monitoring efforts; 2) technology or tool training, in particular for network and system administrators, system developers, firewall administrators, Network Operations Center staff, and incident response professionals; 3) role-specific training, such as training for an ISSO, guiding the activities those professionals need to perform for that role; and 4) training specific to relevant professional certifications, such as the Certified Information Systems Security Professional. The need for this role-based training is summarized in figure 1, which is data collected directly from our survey, sorted by program office or regional office, which illustrates the percentages of EPA personnel with significant information security responsibilities who do not feel they have sufficient role-based training to perform their duties.

Figure 1: Percentage of EPA personnel who believe they do not have sufficient IT security role-based training (by program office and regional location) [1]

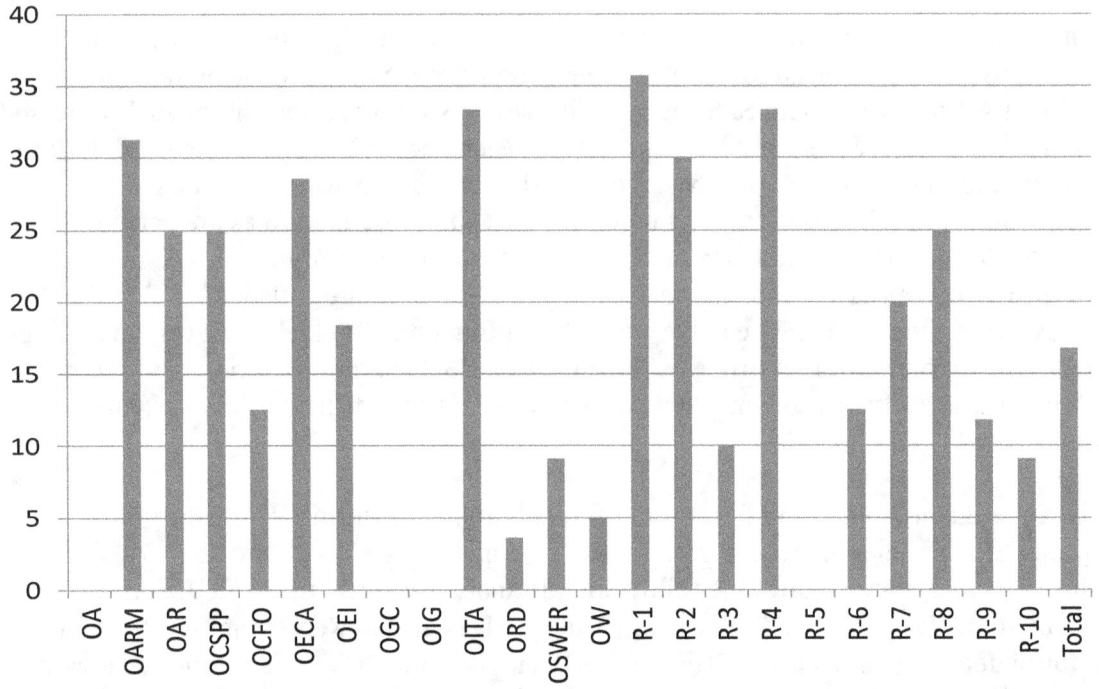

Source: Independent public accounting firm's (IPA's) survey results.

---

[1] Zero percent of Office of General Counsel, Office of Inspector General and Region 5 survey respondents indicated that they did not believe they have sufficient IT security role-based training.

Available Skillport Training Does Not Align With EPA Professionals' Needs

Skillport is the EPA's online training tool, and one of the primary tools for offering information security training to EPA professionals. We noted that Skillport provides essentially the same set of courses for all information security personnel regardless of role. Although the EPA refers to Skillport as a "role-based" training tool, there is limited distinction made in the training requirements between executive and technical personnel. Based on inputs from EPA personnel feedback from the interviews and survey responses, the Skillport training appears to be too technical for executive level personnel and too general for technical personnel. The results for this issue are summarized in figure 2, which is data from our interviews, and illustrates that only 27 percent of the interview participants felt that Skillport offered adequate role-based training. Notably, 42 percent of interviewees felt that security training needed more EPA and role focus. They noted that available training within Skillport lacked content specific to the EPA, the federal environment, and/or the respondents' assigned information security role(s).

Figure 2: Percentage of EPA personnel that receive value from Skillport

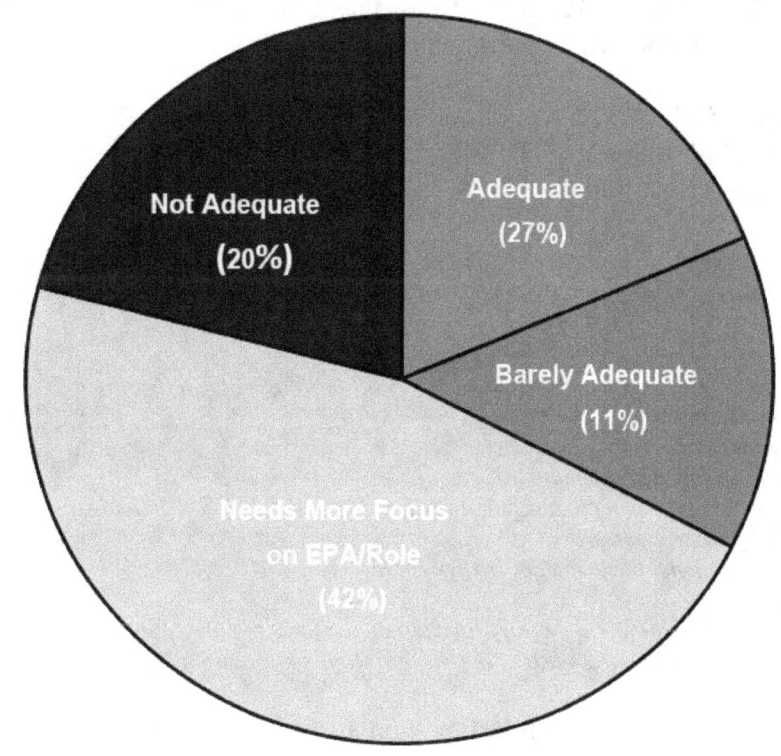

Source: IPA's interviews' results.

In addition, our survey responses noted that 62 (17 percent) of 356 respondents, many of whom were assigned an information security role as a "collateral duty," indicated they did not have adequate experience and training to perform their information security role. Further, the EPA does not provide consistent information security basic training for executive-level personnel who are new to their security role, such as SIOs. Such executives need to have a sound understanding of the FISMA requirements, supporting NIST controls, and corresponding EPA information security policies and procedures, as well as a basic understanding of information security risks and risk management, so they are prepared to make decisions impacting the EPA's information security posture.

The issues we identified during this review are consistent with issues identified during a Government Accountability Office (GAO) report issued in July 2012.[2] In this report, GAO recommended that the EPA develop and finalize a role-based security training program tailored to the specific training requirements of EPA users' role/position descriptions and to the actions ISOs must take when users do not complete the training. The EPA agreed with the GAO recommendation and responded that it would continue analyzing information security roles and responsibilities for personnel with significant security responsibilities and develop and implement a tailored role-based training program. However, at the time of our review we did not see evidence that the EPA has implemented this tailored role-based training program.
We found that these issues exist because the EPA has not completed efforts to clearly align relevant information security training to the specific security-related roles supporting operations within the EPA.

In addition to the E-Government Act, the EPA also needs to comply with the following information security guidelines and requirements:

- EPA CIO Policy 2150.3, *EPA Information Security Policy*, August 6, 2012, specifically EPA Information Procedures CIO 2150.3-P-02.1, *Information Security - Interim Awareness and Training Procedures v3.1*, July 18, 2012:

  EPA shall determine the appropriate content of the security training based on assigned roles and responsibilities and the specific requirements of the information systems to which personnel have authorized access.

  EPA shall provide adequate security-related technical training to the following individuals in order for them to perform their assigned duties:
  - i. *Information system managers.*
  - ii. *System and network administrators.*
  - iii. *Personnel performing independent verification and validation activities.*
  - iv. *Security control assessors.*
  - v. *Other personnel having access to system-level software.*

  EPA shall provide the training necessary for these individuals to carry out their responsibilities related to operations security within the context of the organization's information security program.

- NIST Special Publication (SP) 800-53 Revision 3 *Recommended Security Controls for Federal Information Systems and Organizations,*" August 2009:

  AT-1, Security Awareness and Training Policy and Procedures: The organization develops, documents, and disseminates to [Assignment: organization-defined personnel or roles]:
  - *1. A security awareness and training policy that addresses purpose, scope, roles, responsibilities, management commitment, coordination among organizational entities, and compliance; and*

---

[2] GAO-12-696 *INFORMATION SECURITY- Environmental Protection Agency Needs to Resolve Weaknesses*, July 2012

2. *Procedures to facilitate the implementation of the security awareness and training policy and associated security awareness and training controls.*

AT-3, Role-Based Security Training: The organization provides role-based security training to personnel with assigned security roles and responsibilities:
1. *Before authorizing access to the information system or performing assigned duties;*
2. *When required by information system changes; and*
3. *[Assignment: organization-defined frequency] thereafter.*

- OPM Title 5, Code of Federal Regulations Part 930.301, *Information Systems Security Awareness Training Program*:

  Executives must receive training in information security basics and policy level training in security planning and management.

  Program and functional managers must receive training in information security basics; management and implementation level training in security planning and system/application security management; and management and implementation level training in system/application life cycle management, risk management, and contingency planning.

  CIOs, IT security program managers, auditors, and other security-oriented personnel (e.g., system and network administrators, and system/application security officers) must receive training in information security basics and broad training in security planning, system and application security management, system/application life cycle management, risk management, and contingency planning.

  IT function management and operations personnel must receive training in information security basics; management and implementation level training in security planning and system/application security management; and management and implementation level training in system/application life cycle management, risk management, and contingency planning.

In addition to complying with government guidelines and requirements, the EPA needs to improve in these areas to help ensure that agency personnel have adequate experience and training to perform their assigned information security roles.

Recommendations

KPMG recommends that the Assistant Administrator for Environmental Information:

1. Define key information security aspects and duties for each security role. This includes identifying, where appropriate, broadly similar characteristics within each role to allow for more precise alignment of roles to applicable training requirements. This also includes ensuring that existing EPA policies, procedures, and guidance fully and consistently define all information security roles and responsibilities currently implemented across the organization.

2. Provide additional training options specific to the federal information security environment and EPA information security roles, such as the processes and controls outlined in NIST SP 800-53. Training should be specific to supporting EPA professionals in executing and performing assigned information security roles and responsibilities in accordance with EPA policies and procedures. For example, vendor training may be

warranted for hands-on information security roles, but general orientation training may be suitable for executives.

Agency Comments and KPMG Evaluation

The agency agreed with these recommendations and provided a response to the draft report which included corrective actions and milestone dates. We found the response to be acceptable. Subsequent to issuing the draft report, KPMG and the OIG met with the agency to discuss the report's findings and recommendations. As a result of those discussions and the agency's response to the draft, we updated the report as appropriate.

The agency initially did not agree with draft report recommendation 3. The agency stated further clarification was requested for this recommendation. We subsequently met with agency representatives to discuss the recommendation and updated recommendation 1 to include elements of recommendation 3 relative to the alignment of roles to information security training requirements. The agency concurred with the updated recommendation and provided a high-level corrective action plan with completion dates.

Chapter 3
Information Security Governance Supporting Training Efforts Can Be Improved

Implementation of the EPA's information security training program is hampered by inconsistent assignment of information security roles across the various regional and program offices. These offices' organizational structures vary widely, resulting in differing governance models and, consequently, differing levels of implementation and oversight of training requirements for the EPA personnel performing technical and managerial functions related to information security. These inconsistencies result in inadequate implementation of the EPA, NIST, and OPM requirements related to the provisioning of focused, role-based training for individuals at all levels throughout the EPA's hierarchy. This organizational structure can further expose the organization and its systems and underlying sensitive data to the risk of unauthorized access, misuse or disclosure.

Assignment of Information Security Roles Is Inconsistent

The process for assigning information security roles varies across the EPA offices. For instance, some information security duties are formally defined within position descriptions, while others have information security responsibilities as a collateral duty. Assignment of information security roles to individuals does not necessarily consider whether the individual has sufficient previous relevant experience to ensure the adequacy of security controls for the information system(s) for which they are responsible. We learned that the EPA assigns information security roles to individuals in an informal manner and does not link the roles to established position descriptions or to the agency's Performance Appraisal and Recognition System. Specifically, position descriptions and corresponding Performance Appraisal and Recognition System elements are typically limited to baseline descriptions defined by OPM and are not further tailored to reflect additional "collateral" responsibilities, nor are revised as responsibilities are assigned subsequent to the individual being placed in the position initially. Further, the EPA does not consistently define the various agency information security roles. For instance, it is unclear whether individuals are required to have certain credentials (e.g., professional certifications, in-house training) or experience with specific technologies, platforms, or utilities necessary to implement or monitor technical controls on the EPA's networks and systems.

Organization Structure for Information Security Roles Is Inconsistent

There are different information security governance and organizational models across the EPA. For instance, in some regional and program offices, the SIO has direct oversight of IT operations within the organization, while in others a separate Program Manager is designated with this oversight of IT operations. In some cases an Information Management Officer (IMO) has just an oversight role, while in others the IMO is also the Branch Chief and supervisor for operations personnel. Additionally, a Primary ISO for an office or region may receive guidance from the Senior Agency ISO, SIO and possibly the IMO. In turn, the Primary ISO may give guidance to System Owners, ISSOs, local ISOs and alternates. A System Owner may have to respond to the

Primary ISO, Program Manager, ISO and IMO. Consequently, the EPA has not consistently established lines of authority and the expected interaction between various information security roles. This inconsistency greatly complicates the tasks of defining necessary skills and identifying which organizational level should have the responsibility for providing the appropriate type of role-based training.

We found that these issues exist because the EPA has a decentralized information security management structure, with responsibilities shared among many organizational components, including headquarters, regional offices, and the National Computer Center in Research Triangle Park. Although there is one SAISO with EPA-wide responsibilities, including coordination with the Office of Environmental Information's (OEI's) Mission Investment Solutions Division for the development and implementation of EPA's agencywide information security training program, there are also 23 Primary ISOs with similar roles in each office and region. Further, the EPA has not made readily apparent the extent of responsibility and cooperation needed among these organizations for ensuring that information security personnel in all roles have the necessary skills and training. Finally, some information security roles are collateral duties held by personnel whose position and primary work responsibilities may entail unrelated functions. Consequently, the individual assigned these types of collateral information security roles may have little or no experience. Such roles include SIO, IMO, Program Manager, System Owner, and Contracting Officer Representative. OEI security management should take on a more prominent role in ensuring that EPA information security personnel complete necessary training, as in some cases the ISOs do not have the authority to ensure personnel comply with the training requirements.

In addition to the E-Government Act, the EPA also needs to comply with the following information security guidelines and requirements:

- EPA CIO Policy 2150.3, *EPA Information Security Policy*, August 6, 2012, specifically P-02.,1 *Information Security – Interim Awareness and Training Procedures,* v3.1, July 18, 2012:

> *EPA shall determine the appropriate content of the security training based on assigned roles and responsibilities and the specific requirements of the information systems to which personnel have authorized access.*
>
> *EPA shall provide adequate security-related technical training to the following individuals in order for them to perform their assigned duties:*
> i. *Information system managers.*
> ii. *System and network administrators.*
> iii. *Personnel performing independent verification and validation activities.*
> iv. *Security control assessors.*
> v. *Other personnel having access to system-level software.*
>
> *EPA shall provide the training necessary for these individuals to carry out their responsibilities related to operations security within the context of the organization's information security program.*

- EPA CIO Policy 05-001, *Senior Information Officials*, July 7, 2005:

    Due to increasing legal requirements and good management practices, information and information technology management responsibilities and functions are becoming more important for accomplishing the Agency's mission, and the scope and complexity of those responsibilities and functions continue to expand. To ensure these increasingly more important responsibilities and functions are performed effectively throughout EPA, the Agency's organizations must have appropriate accountability for this critical area. A designated Senior Information Official in each Program and Regional Office will ensure EPA's information and information technology are effectively managed both corporately across the Agency and within each organization to achieve EPA's business needs, mission, and strategic goals, and will help the Agency achieve a cohesive, comprehensive approach to its information and information technology infrastructure, architecture, security, web policies, and public access.

- NIST SP 800-53 Revision 3, *Recommended Security Controls for Federal Information Systems and Organizations*, August 2009:

    AT-1, Security Awareness and Training Policy and Procedures: The organization develops, documents, and disseminates to [Assignment: organization-defined personnel or roles]:
    3. *A security awareness and training policy that addresses purpose, scope, roles, responsibilities, management commitment, coordination among organizational entities, and compliance; and*
    4. *Procedures to facilitate the implementation of the security awareness and training policy and associated security awareness and training controls.*

    AT-3, Role-Based Security Training: The organization provides role-based security training to personnel with assigned security roles and responsibilities:
    4. *Before authorizing access to the information system or performing assigned duties;*
    5. *When required by information system changes; and*
    6. *[Assignment: organization-defined frequency] thereafter.*

- OPM 5 CFR Part 930.301, *Information Systems Security Awareness Training Program:*

    Executives must receive training in information security basics and policy level training in security planning and management.

    Program and functional managers must receive training in information security basics; management and implementation level training in security planning and system/application security management; and management and implementation level training in system/application life cycle management, risk management, and contingency planning.

    CIOs, IT security program managers, auditors, and other security-oriented personnel (e.g., system and network administrators, and system/application security officers) must receive training in information security basics and broad training in security planning, system and application security management, system/application life cycle management, risk management, and contingency planning.

    IT function management and operations personnel must receive training in information security basics; management and implementation level training in security planning and system/application security management; and management and implementation level training in system/application life cycle management, risk management, and contingency planning.

In addition to complying with government guidelines and requirements, the EPA needs to improve these areas because the lack of a centralized governance structure leads to inconsistencies in the operation of the information security training program.

Recommendations

KPMG recommends that the Assistant Administrator for Environmental Information:

3. Standardize the terminology and definition of responsibilities for key IT security management and oversight roles across all EPA organizations and within the EPA information security policy.

4. Provide a more clear delineation of which EPA organizations should be responsible for delivering specific elements of information security role training, and how collectively and cooperatively the training needs of each significant role (including technical and executive-level roles) are to be met.

Agency Comments and KPMG Evaluation

The agency agreed with these recommendations and provided us with a response to the draft report which included corrective actions with milestone dates. We found the response to be acceptable and updated the report as appropriate.

Status of Recommendations and Potential Monetary Benefits

		RECOMMENDATIONS				POTENTIAL MONETARY BENEFITS (in $000s)	
Rec. No.	Page No.	Subject	Status[1]	Action Official	Planned Completion Date	Claimed Amount	Agreed-To Amount
1	7	Define key information security aspects and duties for each security role. This includes identifying, where appropriate, broadly similar characteristics within each role to allow for more precise alignment of roles to applicable training requirements. This also includes ensuring that existing EPA policies, procedures, and guidance fully and consistently define all information security roles and responsibilities currently implemented across the organization.	O	Assistant Administrator for Environmental Information	12/31/2016		
2	7	Provide additional training options specific to the federal information security environment and EPA information security roles, such as the processes and controls outlined in NIST SP 800-53. Training should be specific to supporting EPA professionals in executing and performing assigned information security roles and responsibilities in accordance with EPA policies and procedures. For example, vendor training may be warranted for hands-on information security roles, but general orientation training may be suitable for executives.	O	Assistant Administrator for Environmental Information	12/31/2016		
3	12	Standardize the terminology and definition of responsibilities for key IT security management and oversight roles across all EPA organizations and within the EPA information security policy.	O	Assistant Administrator for Environmental Information	09/30/2015		
4	12	Provide a more clear delineation of which EPA organizations should be responsible for delivering specific elements of information security role training, and how collectively and cooperatively the training needs of each significant role (including technical and executive-level roles) are to be met.	O	Assistant Administrator for Environmental Information	12/31/2015		

[1] O = recommendation is open with agreed-to corrective actions pending
   C = recommendation is closed with all agreed-to actions completed
   U = recommendation is unresolved with resolution efforts in progress

Survey Questions

Note that the survey was Web based and each question had multiple responses available for the respondent to select. At the end of the survey, we also included several optional questions for the respondent to consider.

1. In which EPA office are you currently performing IT security duties?

2. Select one or more of the following roles that <u>most closely </u>describe your IT security-related function(s) within the Agency.

3. For each of the roles you selected in the previous question, please specify the approximate percentage of time you spend performing functions associated with that role. To account for any non-IT security-related roles that you may hold additionally, ensure that the total, including percentage of time spent on "non-IT security related roles" equals 100.

4. Were you required to participate in specialized training <u>prior to </u>assuming your EPA IT security responsibilities?

5. How frequently do you attend or participate in specialized training (including "refresher" training) associated with one or more IT security functions for which you have management or operational responsibility over?

6. Is your participation in specialized training mandated by your supervisor or other EPA management?

7. How many years of IT security-related experience did you possess <u>prior to </u>assuming responsibility for that function at EPA?

8. Which of the following IT security functional areas do you have management or operational responsibility over (check as many as apply) (example responses included Data Security, Digital Forensics, IT Security Training, etc.).

9. For each of the functional areas you selected in the previous question, please specify the approximate percentage of time you spend performing that function <u>as a component of your overall IT security responsibilities at EPA</u>. The total of all values should equal 100 percent of time spent on IT security responsibilities. Please enter whole numbers, no percentage signs.

10. Rate your level of experience and knowledge in your role that is associated with each of the following functional areas of IT security using a scale of 1 to 5 as defined below.

11. What IT security-related certifications do you hold?

12. Please select all levels of post-high school education attained.

13. Please select the degree(s) program, major, or area of study completed. You may select more than 1 degree if applicable.

14. Within the last three years, for which of the following IT security functional areas have you received specialized or targeted job related training? Place your cursor over the answer choices to see a short description. Example responses include Data Security, Telecommunications Security, Personnel Security, etc.

15. For Data Security, please select the terms and concepts associated with specialized or targeted job related training that you have received within the past three years. Example responses include Access Controls, Antivirus Software, Authentication, etc.

16. Do you believe you have the training and prior experience needed to effectively and efficiently perform your assigned IT security related responsibilities?

17. For which specific areas or functions do you feel additional training and/or experience would better enable you to effectively and efficiently perform your assigned IT security related responsibilities?

18. How familiar are you with federal and/or EPA policies or requirements pertaining to specialized training for IT security roles and responsibilities?

19. What factors (if any) hinder you in performing your IT security duties (e.g., undue influence from your supervisor or senior personnel, lack of authority to provide direction, lack of training, etc.)?

20. In your opinion, how can EPA strengthen its IT security program?

21. Would you like someone from KPMG to contact you to confidentially discuss your survey responses in greater detail?

22. Optional Question -Answer this question at your discretion.
Within the last three years, have you attended any specialized or targeted job related training courses that were particularly or significantly beneficial in improving your capabilities to support EPA's IT security program?

23. Optional Question -Answer this question at your discretion.
Please be as specific as possible in identifying the course title, source, subject matter and corresponding capabilities improved by your attendance.

24. Optional Question -Answer this question at your discretion.
Within the last three years, were there any job related training courses that did NOT improve your capabilities?

25. Optional Question -Answer this question at your discretion.
Please be as specific as possible in identifying the course title, subject matter and intended capabilities that were not addressed by the course.

26. Optional Question -Answer this question at your discretion.
Please provide any additional general comments or feedback concerning EPA's IT security program, training and development requirements and resources, or other relevant topic areas.

Agency Response to Draft Report

<u>**MEMORANDUM**</u>

SUBJECT: Response to Office of Inspector General Draft Report No. OMS-FY12-0006 "EPA Should Enhance Existing Training Program for Personnel with Significant Information Security Responsibilities," dated January 16, 2014

FROM: Reneé P. Wynn /s/
                 Acting Assistant Administrator and Acting Chief Information Officer

TO: Arthur A. Elkins, Jr.
                 Inspector General

Thank you for the opportunity to respond to the issues and recommendations in the subject audit report. Following is a summary of the agency's overall position, along with its position on each of the report recommendations. For those report recommendations with which the agency agrees, we have provided high-level intended corrective actions and estimated completion dates. For the recommendation with which the OEI does not agree, we have explained our position, and proposed alternatives to recommendations.

AGENCY'S OVERALL POSITION
Of the five recommendations in the draft audit report, OEI agrees with 1, 2, 4 and 5 and describes high-level intended corrective actions in the attached table.

SUMMARY OF DISAGREEMENTS
With respect to recommendation 3, OEI disagrees because further clarification is requested for this recommendation.

If you have any questions regarding this response, please contact Robert McKinney, subject audit primary contact, Senior Agency Information Security Officer (SAISO), at (202) 564-0921, mckinney.robert@epa.gov or Scott Dockum, OEI Audit Follow-Up Manager, Office of Program Management, Policy, Outreach and Communications Staff at (202) 566-1914, dockum.scott@epa.gov.

Attachment

AGENCY'S RESPONSE TO REPORT RECOMMENDATIONS

Agreements

No.	Recommendation	High-Level Intended Corrective Action(s)	Estimated Completion by Quarter and FY
1	Define key information security aspects and duties for each security role to allow for more defined training and ensure that existing EPA policies, procedures, and guidance fully and consistently define all information security roles and responsibilities currently implemented across the organization.	OEI will continue to refine identified roles and their respective responsibilities in the agency Roles and Responsibilities procedure (CIO-215-.3-P-19.1), the reference document for information security roles and responsibilities. OEI will ensure role names are consistently used throughout OEI developed policies, procedures and guidelines.	Quarter 4, FY15
2	Provide additional training options specific to the federal information security environment and EPA information security roles, such as the processes and controls outlined in NIST SP 800-53. Training should be specific to supporting EPA professionals in executing and performing assigned information security roles and responsibilities in accordance with EPA policies, and procedures. For example, vendor training may be warranted for hands on information security roles, but general orientation training may be suitable for executives.	OEI will review training options and inform agency personnel of appropriate training options for each identified role(s).	Quarter 1, FY16

No.	Recommendation	High-Level Intended Corrective Action(s)	Estimated Completion by Quarter and FY
4	Standardize the terminology and definition of responsibilities for key IT security management and oversight roles across all EPA organizations and within the EPA information security policy.	OEI will continue to support the consistent use of terminology and definitions for key IT security roles. OEI will continue to refine and update the roles and responsibilities procedure, CIO-215-.3-P-19.1, as necessary.	Quarter 4, FY15
5	Provide a more clear delineation of which EPA organizations should be responsible for delivering specific elements of information security role training, and how collectively and cooperatively the training needs of each significant role (including technical and executive level roles) are to be met.	OEI is developing a role based training program that addresses training requirements for both technical and non-technical roles that have significant information security responsibilities.	Quarter 1, FY15

Disagreements

No.	Recommendation	Agency Explanation/Response	Proposed Alternative
3	Complement the Skillport training process by establishing an appropriate number of information security roles that identify broadly similar characteristics and inherently governmental roles, and link these roles to applicable training requirements.	Further clarification is requested for this recommendation.	N/A

Revised Agency Corrective Action Plan
to Report Recommendations

No.	Recommendation	High-Level Intended Corrective Action(s)	Estimated Completion by Quarter and FY
1	Define key information security aspects and duties for each security role. This includes identifying, where appropriate, broadly similar characteristics within each role to allow for more precise alignment of roles to applicable training requirements. This also includes ensuring that existing EPA policies, procedures, and guidance fully and consistently define all information security roles and responsibilities currently implemented across the organization	In developing a role based training program, OEI will define the responsibilities for each role and closely align them to appropriate training. OEI will continue to develop new and update existing policies, procedures, and guidance under OEI's purview so that information security roles and responsibilities are defined consistently.	Quarter 1, FY16
2	Provide additional training options specific to the federal information security environment and EPA information security roles, such as the processes and controls outlined in NIST SP 800-53. Training should be specific to supporting EPA professionals in executing and performing assigned information security roles and responsibilities in accordance with EPA policies, and procedures. For example, vendor training may be warranted for hands on information security roles, but general orientation training may be suitable for executives.	OEI will review training options and inform agency personnel of appropriate training options for each identified role(s).	Quarter 1, FY16

No.	Recommendation	High-Level Intended Corrective Action(s)	Estimated Completion by Quarter and FY
3	Standardize the terminology and definition of responsibilities for key IT security management and oversight roles across all EPA organizations and within the EPA information security policy.	OEI will continue to support the consistent use of terminology and definitions for key IT security roles. OEI will continue to refine and update the roles and responsibilities procedure, CIO-215-.3-P-19.1, as necessary.	Quarter 4, FY15
4	Provide a more clear delineation of which EPA organizations should be responsible for delivering specific elements of information security role training, and how collectively and cooperatively the training needs of each significant role (including technical and executive level roles) are to be met.	OEI is developing a role based training program that addresses training requirements for both technical and non-technical roles that have significant information security responsibilities.	Quarter 1, FY15

Distribution

Office of the Administrator
Assistant Administrator for Environmental Information and Chief Information Officer
Agency Follow-Up Official (the CFO)
Agency Follow-Up Coordinator
General Counsel
Associate Administrator for Congressional and Intergovernmental Relations
Associate Administrator for External Affairs and Environmental Education
Senior Agency Information Security Officer
Audit Follow-Up Coordinator, Office of Environmental Information

www.ingramcontent.com/pod-product-compliance
Lightning Source LLC
Chambersburg PA
CBHW081247170526
45165CB00009B/3227